Andrew Ross

The Microscope

Andrew Ross

The Microscope

ISBN/EAN: 9783337075460

Printed in Europe, USA, Canada, Australia, Japan

Cover: Foto ©berggeist007 / pixelio.de

More available books at **www.hansebooks.com**

THE

MICROSCOPE.

BEING THE ARTICLE CONTRIBUTED BY

ANDREW ROSS

TO THE "PENNY CYCLOPÆDIA," PUBLISHED BY THE SOCIETY

FOR THE DIFFUSION OF USEFUL KNOWLEDGE.

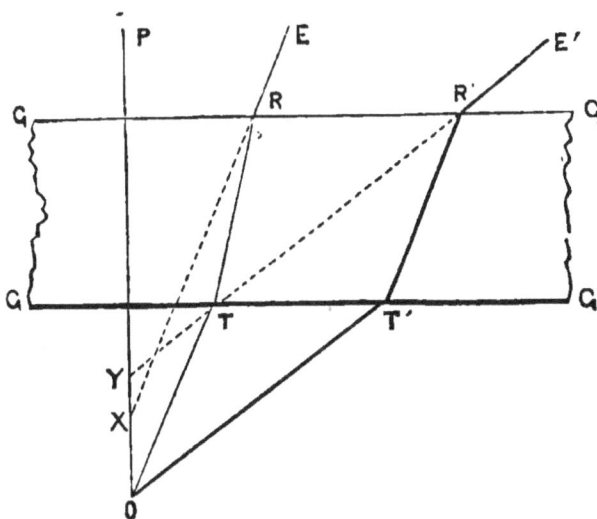

FULLY ILLUSTRATED.

NEW YORK:

THE INDUSTRIAL PUBLICATION COMPANY.

1877.

THE MICROSCOPE.

———— • ✦ • ————

ICROSCOPE, the name of an instrument for enabling the eye to see distinctly objects which are placed at a very short distance from it, or to see magnified images of small objects, and therefore to see smaller objects than would otherwise be visible. The name is derived from the two Greek words, expressing this property, MIKROS, *small*, and SKOPEO, *to see*.

So little is known of the early history of the microscope, and so certain is it that the magnifying power of lenses must have been discovered as soon as lenses were made, that there is no reason for hazarding any doubtful speculations on the question of discovery. We shall proceed therefore at once to describe the simplest forms of microscopes, to explain their later and more important improvements, and finally to exhibit the instrument in its present perfect state.

In doing this we shall assume that the reader is familiar with the information contained in the articles "Light." "Lens," "Achromatic," "Aberration," and the other sub-divisions of the science of Optics, which are treated of in this work.

The use of the term *magnifying* has led many into a misconception of the nature of the effect produced by convex lenses. It is not always understood that the so-called magnifying power of a lens applied to the eye, as in a microscope, is derived from its enabling the eye to approach more nearly to its object than would otherwise be compatible with distinct vision. The common occurrence of walking across the street to read a bill is in fact magnifying the bill by approach; and the observer, at every step he takes, makes a change in the optical arrangement of his eye, to adapt it to the lessening distance between himself and the object of his inquiry. This power of spontaneous adjustment is so unconsciously exerted, that unless the attention be called to it by circumstances, we are totally unaware of its exercise.

In the case just mentioned the bill would be read with eyes in a very different state of adjustment from that in which it was discovered on the opposite side of the street, but no conviction of this fact would be impressed upon the mind. If, however, the supposed individual should perceive on some part of the paper a small speck, which he suspects to be a minute insect, and if he should attempt a very close approach of his eye for the purpose of verifying his suspicion, he would presently find that the power of natural adjustment has a limit; for when his eye has arrived within about ten inches, he will discover that a further approach produces only confusion. But if, as he continues to approach, he were to place before his eye a series of properly arranged convex lenses, he would see the object gradually and distinctly increase in apparent size by the mere continuance of the operation of approaching. Yet the glasses applied to the eye during the approach from ten inches to one inch, would have done nothing more than had been previously done by the eye itself during the approach from fifty feet to one foot. In both cases the magnifying is effected really by the approach, the lenses merely rendering the latter periods of the approach compatible with distinct vision.

'A very striking proof of this statement may be obtained by the following simple and instructive experiment. Take any minute object, a very small insect for instance, held on a pin or gummed to a slip of glass; then present it to a strong light, and look at it through the finest needle-hole in a blackened card placed about an inch before it. The insect will appear quite distinct, and about ten times larger than its usual size. Then suddenly withdraw the card without disturbing the object, which will instantly become indistinct and nearly invisible. The reason is, that the naked eye cannot see at so small a distance as one inch. But the card with the hole having enabled the eye to approach within an inch, and to see distinctly at that distance, is thus proved to be as decidedly a magnifying instrument as any lens or combination of lenses.

This description of magnifying power does not apply to such instruments as the solar or gas microscope, by which we look not at the object itself, but at its shadow or picture on the wall; and the description will require some modification in treating of the compound microscope, where, as in the telescope, an image or picture is formed by one lens, that image or picture being viewed as an original object by another lens.

It is nevertheless so important to obtain a clear notion of the real nature of the effect produced by a lens applied to the eye, that we will adduce the instance of spectacles to render the point more familiar. If the person who has been supposed to cross the street for the purpose of reading a bill had been aged, the limit to the power of adjustment would have been discovered at a greater distance, and without so severe a test as the supposed insect. The eyes of the very aged generally lose the power of adjustment at a distance of thirty or forty inches instead of ten, and the spectacles worn in consequence are as much magnifying glasses to them as the lenses employed by younger eyes to examine the most minute objects. Spectacles are magnifying glasses to the aged because they enable such persons to see as closely to their objects as the young, and therefore to see the objects larger than they could themselves otherwise see them, but not larger than they are seen by the unassisted younger eye.

In saying that an object appears larger at one time, or to one person, than another, it is necessary to guard against misconception. By the apparent size of an object we mean the angle it subtends at the eye, or the angle formed by two lines drawn from the centre of the eye to the extremities of the object. In Fig. 1, the lines A E and B E drawn from the arrow to the eye form the angle A E B, which, when the angle is small, is nearly twice as great as the angle C E D, formed by lines drawn from a similar arrow at twice the distance. The arrow A B will therefore appear nearly twice as long as C D, being seen under twice the angle, and in the same proportion for any greater or lesser difference in distance. The angle in question is called the angle of vision, or the visual angle.

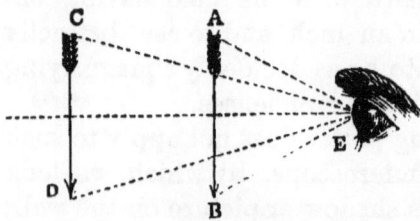

The angle of vision must, however, not be confounded with the angle of the pencil of light by which an object is seen, and which is explained in Fig. 2. Here we have drawn two arrows placed in relation to the eye as before, and from the centre of each have drawn lines exhibiting the quantity of light which each point will send into the eye at the respective distances.

Fig. 1.

Now if E F represent the diameter of the pupil, the angle E A F shows the size of the cone or pencil of light which enters the eye from the point A, and in like manner the angle E B F is that of the pencil emanating from B, and entering the eye. Then, since E A F is double E B F, it is evident that A is seen by four times the quantity of light which could be received from an equally illuminated point at B; so that the nearer body would appear brighter if

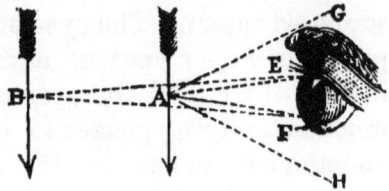

Fig. 2.

it did not appear larger; but as its apparent area is increased four times as well as its light, no difference in this respect is discovered. But if we could find means to send into the eye a larger pencil of light, as for instance that shown by the lines G A H, without increasing the apparent size in the same proportion, it is evident that we should obtain a benefit totally distinct from that of increased magnitude, and one which is in some cases of even more importance than size in developing the structure of what we wish to examine. This, it will be hereafter shown, is sometimes done; for the present, we wish merely to explain clearly the distinction between apparent magnitude, or the angle under which the object is seen, and apparent brightness, or the angle of the pencil of light by which each of its points is seen, and with these explanations we shall continue to employ the common expressions magnifying glass and magnifying power.

The magnifying power of a single lens depends upon its focal length, the object being in fact placed nearly in its principal focus, or so that the light which diverges from each point may, after refraction by the lens, proceed in parallel lines to the eye, or as nearly so as is requisite for distinct vision. In Fig. 3, A B is a double convex lens, near which is a small arrow to

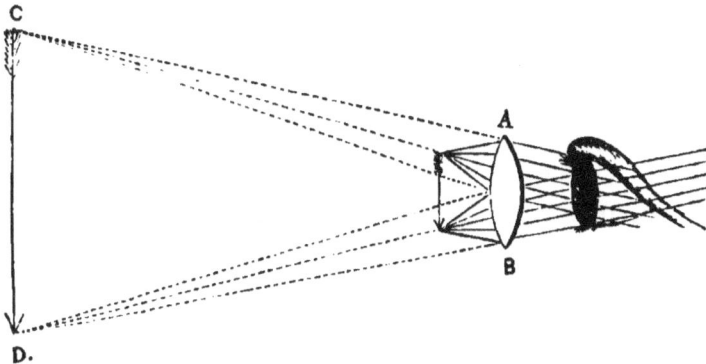

Fig. 3.

represent the object under examination, and the cones drawn from its extremities are portions of the rays of light diverging from those points and falling upon the lens. These rays, if suffered to fall at once upon the pupil, would be too divergent to permit their being brought to a focus upon the retina by the optical arrangements of the eye. But being first passed through the lens, they are bent into nearly parallel lines, or into lines diverging from some points within the limits of distinct vision, as from C and D. Thus altered, the eye receives them precisely as if they emanated from a larger arrow placed at C D, which we may suppose to be ten inches from the eye, and then the difference between the real and the imaginary arrow is called the magnifying power of the lens in question.

From what has been said it will be evident that two persons whose eyes differed as to the distance at which they obtained distinct vision, would give different results as to the magnifying power of a lens. To one who can see distinctly with the naked eye at a distance of five inches, the magnifying power would seem and would indeed be only half what we have assumed. Such instances are, however, rare; the focal length of the eye usually ranges from six to twelve or fourteen inches, so that the distance we first assumed of ten inches is very near the true average, and is a convenient number, inasmuch as a cipher added to the denominator of the fraction which expresses the focal length of a lens gives its magnifying power. Thus a lens whose focal length is one-sixteenth of an inch is said to magnify 160 times.

When the focal length of a lens is very small, it is difficult to measure accurately the distance between its centre and its object. In such cases the best way to obtain the focal length for parallel or nearly parallel rays is to view the image of some distant object formed by the lens in question through another lens of one inch solar focal length, keeping both eyes open and comparing the image presented through the two lenses with that of the naked eye. The proportion between the two images so seen will be the focal length required. Thus if the image seen by the naked eye is ten times as large as that shown by the lenses, the focal length of the lens in question is one-tenth of an inch. The panes of glass in a window, or courses of bricks in a wall, are convenient objects for this purpose.

In whichever way the focal length of the lens is ascertained, the rules given for deducing its magnifying power are not rigorously correct, though they are sufficiently so for all practical purposes, particularly as the whole rests on an assumption in regard to the focal length of the eye, and as it does not in any way affect the actual measurement of the object. To calculate with great precision the magnifying power of a lens with a given focal length of eye, it is necessary that the thickness of the lens be taken into the account, and also the focal length of the eye itself.

We have hitherto considered a magnifying lens only in reference to its enlargement of the object, or the increase of the angle under which the object is seen. A further and equally important consideration is that of the number of rays or quantity of light by which every point of the object is rendered visible. The naked eye, as shown in Fig. 2, admits from each point of every visible object a cone of light having the diameter of the pupil for its base, and most persons are familiar with that beautiful provision by which in cases of excessive brilliancy the pupil spontaneously contracts to reduce the cone of admitted light within bearable limits. This effect is still further produced in the experiment already described, of looking at an object through a needle-hole in a card, which is equivalent to reducing the pupil to the size of a needle-hole. Seen in this way the object becomes comparatively dark or obscure; because each point is seen by means of a very small cone of light, and a little consideration will suffice to explain the different effects produced by the needle-hole and the lens. Both change the angular value of the cone of light presented to the eye, but the lens changes the angle by bending the extreme rays within the limits suited to distinct vision, while the needle-hole effects the same purpose by cutting off the rays which exceed those limits.

It has been shown that removing a brilliant object to a greater distance will reduce the quantity of light which each point sends into the eye, as effectually as viewing it through a needle-hole; and magnifying an object by a lens has been shown to be the same thing in some respects as removing it to a greater distance. We have to see the magnified picture by the light emanating from the small object, and it becomes a matter of difficulty to obtain from each point a sufficient quantity of light to bear the diffusion of a great magnifying power. We want to perform an operation just the reverse of applying the card with the needle-hole to the eye—we want in some cases to bring into the eye the largest possible pencil of light from each point of the object.

Referring to Fig. 3, it will be observed that if the eye could see the small arrow at the distance there shown without the intervention of the lens, only a very small portion of the cones of light drawn from its extremities would enter the pupil; whereas we have supposed that after being bent by the lens the whole of this light enters the eye as part of the cones of smaller angle whose summits are at C and D. These cones will further explain the difference between large and small pencils of light; those from the small arrow are large pencils; the dotted cones from the large arrow are small pencils.

In assuming that the whole of this light could have been suffered to enter the eye through the lens A B, we did so for the sake of not perplexing the reader with too many considerations at once. He must now learn that so large a pencil of light passing through a single lens would be so distorted by the spherical figure of the lens, and by the chromatic dispersion of the glass, as to produce a very confused and imperfect image. This confusion may be greatly diminished by reducing the pencil; for instance, by applying a stop, as it is called, to the lens, which is neither more nor less than the needle-hole applied to the eye. A small pencil of light may be thus transmitted through a single lens without suffering from spherical aberration or chromatic dispersion any amount of distortion which will materially affect the figure of the object; but this quantity of light is insufficient to bear diffusion over the magnified picture, which is therefore too obscure to exhibit what we most desire to see—those beautiful and delicate markings by which one kind of organic matter is distinguished from another. With a small aperture these markings are not seen at all: with a large aperture and a single lens they exhibit a faint nebulous appearance enveloped in a chromatic mist, a state which is of course utterly valueless to the naturalist, and not even amusing to the amateur.

It becomes therefore a most important problem to reconcile a large aperture with distinctness, or, as it is called, *denfinition;* and this has been done in a considerable degree by effecting the required amount of refraction through two or more lenses instead of one, thus reducing the angles of incidence and refraction, and producing other effects which will be shortly noticed. This was first accomplished in a satisfactory manner by—

DR. WOLLASTON'S DOUBLET,

invented by the celebrated philosopher whose name it bears; it consists of two plano-convex lenses (Fig. 4) having their focal lengths in the proportion of 1 to 3, or nearly so, and placed at a distance which can be ascertained best by actual experiment. Their plane sides are placed towards the object, and the lens of shortest focal length next the object.

It appears that Dr. Wollaston was led to this invention by considering that the Achromatic Huyghenean Eye-piece, which will be hereafter described, would, if reversed, possess similar good properties as a simple microscope. But it will be evident when the eye-piece is understood, that the circumstances which render it achromatic are very imperfectly applicable to the simple microscope, and that the doublet, without a nice adjustment of the stop, would be valueless. Dr. Wollaston makes no allusion to a stop, nor is it certain that he contemplated its introduction, although his illness, which terminated fatally soon after the presentation of his paper, may account for the omission.

Fig. 4.

The nature of the corrections which take place in the doublet is explained in the annexed diagram (Fig. 5), where L O L′ is the object, P a portion of the pupil, and D D the stop, or limiting aperture.

Now, it will be observed that each of the pencils of light from the extremities L L′ of the object is rendered eccentrical by the stop, and of consequence each passes through the two lenses on opposite sides of their common axis O P; thus each becomes affected by opposite errors, which to some extent balance and correct each other. To take the pencil L, for instance, which enters the eye at R B, R B; it is bent to the right at the first lens, and to the left at the second; and as each bending alters the direction of the blue rays more than the red, and, moreover, as the blue rays fall nearer the margin of the second lens, where the refraction, being more powerful than near the centre, compensates in some degree for the greater focal length of the second lens, the blue and red rays will emerge very nearly parallel, and of consequence colorless to the eye. At the same time the spherical aberration has been diminished by the circumstance that the side of the pencil which passes one lens nearest the axis passes the other nearest the margin.

This explanation applies only to the pencils near the extrem-

ities of the object. The central pencil, it is obvious, would
pass both\lenses symmetrically; the same portions of light
occupying nearly the same relative places on both lenses. The
blue light would enter the second lens nearer to its axis than
the red, and being thus less refracted than the red by the

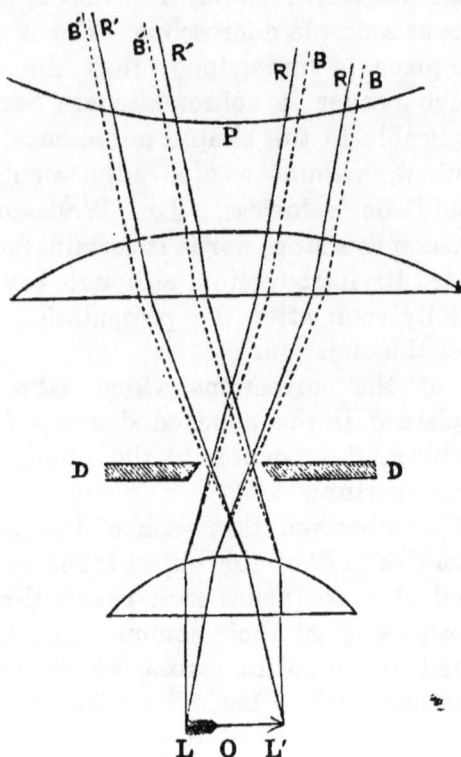

Fig. 5.

second lens, a small amount of compensation would take place,
quite different in principle and inferior in degree to that which
is produced in the eccentrical pencils. In the intermediate
spaces the corrections are still more imperfect and uncertain;
and this explains the cause of the aberrations which must of
necessity exist even in the best-made doublet. It is, however,
infinitely superior to a single lens, and will transmit a pencil of
an angle of from 35° to 50° without any very sensible errors.

It exhibits, therefore, many of the usual test-objects in a very beautiful manner.

The next step in the improvement of the simple microscope bears more analogy to the eye-piece. This improvement was made by Mr. Holland, and it consists (as shown in Fig. 6) in substituting two lenses for the first in the doublet, and retaining the stop between them and the third. The first bending, being thus effected by two lenses instead of one, is accompanied by smaller aberrations, which are therefore more completely balanced or corrected at the second bending, in the opposite direction, by the third lens. This combination, though called a triplet is essentially a doublet, in which the anterior lens is divided into two. For it must be recollected that the first pair of lenses merely accomplishes what might have been done, though with less precision, by one; but the two lenses of the doublet are opposed to each other; the second diminishing the magnifying power of the first. The first pair of lenses in the triplet concur in producing a certain amount of magnifying power, which is diminished in quantity and corrected as to aberration at the third lens by the change in relation to the position of the axis which takes place in the pencil between what is virtually the first and second lens. In this combination the errors are still further reduced by the close approximation to the object which causes the refractions to take place near the axis. Thus the transmission of a still larger angular pencil, namely 65°, is rendered compatible with distinctness, and a more intense image is presented to the eye.

Fig. 6.

Every increase in the number of lenses is attended with one drawback, from the circumstance that a certain portion of light is lost by reflection and absorption each time that the ray enters a new medium. This loss bears no sensible proportion to the gain arising from the increased aperture, which, being as the square of the diameter, multiplies rapidly; or, if we estimate by the angle of the admitted pencil, which is more easily ascertained, the intensity will be as the square of twice the tangent of half the angle. To explain this, let D B (Fig. 7) represent the diameter of the lens, or of that part of it which is really employed; C A the perpendicular drawn from its centre, and A B, A D, the extreme rays of the incident pencil of light D A B. Then the diameter being 2 C B, the area to which the intensity of vision is proportional will be $(2 C B)^2$, and C B is evidently the tangent of the angle C A B, which is half the angle of the admitted pencil D A B. Or, if a be used to denote the angular aperture, the expression for the intensity is $(2 \tan. \frac{1}{2}a)^2$, which increases so rapidly with the increase of a as to make the loss of light by reflection and absorption of little consequence.

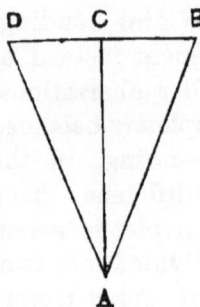

Fig. 7.

The combination of three lenses approaches, as has been stated, very close to the object; so close, indeed, as to prevent the use of more than three; and this constitutes a limit to the improvement of the simple microscope, for it is called a simple microscope, although consisting of three lenses, and although a compound microscope may be made of only three or even two lenses; but the different arrangement which gives rise to the term compound will be better understood when that instrument is explained.

Before we proceed to describe the simple microscope and its appendages, it will be well to explain such other points in reference to the form and materials of lenses as are most likely to be interesting.

A very useful form of lens was proposed by Dr. Wollaston, and called by him the Periscopic lens. It consisted of two hemispherical lenses, cemented together by their plane faces, having a stop between them to limit the aperture. A similar proposal was made by Mr. Coddington, who, however, executed the project in a better manner, by cutting a groove in a whole sphere, and filling the groove with opaque matter. His lens, which is the well-known Coddington lens, is shown in Fig. 8. It gives a large field of view, which is equally good in all directions, as it is evident that the pencils A A and B B pass through under precisely the same circumstances. Its spherical form has the further advantage of rendering the position in which it is held of comparatively little consequence. It is therefore very convenient as a hand-lens, but its definition is of course not so good as that of a well-made doublet or achromatic lens.

Another very useful form of doublet was proposed by Sir John Herschel, chiefly like the Coddington lens, for the sake of a wide field, and chiefly to be used in the hand. It is shown in Fig. 9; it consists of a double convex or crossed lens, having the radii of curvature as 1 to 6, and of a plane concave lens whose focal length is to that of the convex lens as 13 to 5.

Various, indeed innumerable, other forms and combinations of lenses have been projected, some displaying much ingenuity, but few of any practical use. In the Catadioptric lenses the light emerges at right angles from its entering direction, being reflected from a surface cut at an angle of 45 degrees to the axes of the curved surfaces.

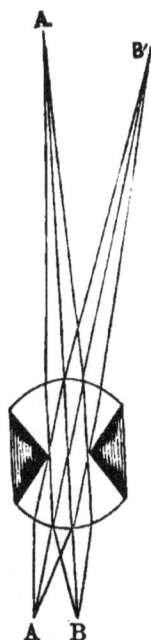

Fig. 8.

It was at one time hoped, as the precious stones are more refractive than glass, and as the increased refractive power is unaccompanied by a correspondent increase in chromatic dispersion, that they would furnish valuable materials for lenses, inasmuch as the refractions would be accomplished by shallower curves, and consequently with diminished spherical aberration. But these hopes were disappointed; everything that ingenuity and perseverance could accomplish was tried by Mr. Varley and Mr. Pritchard, under the patronage of Dr. Goring. It appeared, however, that the great reflective power, the doubly-refracting property, the color, and the heterogeneous structure of the jewels which were tried, much more than

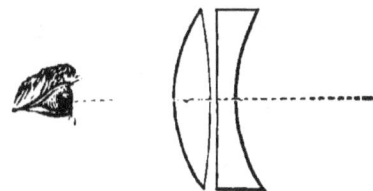

Fig. 9.

counterbalanced the benefits arising from their greater refractive power, and left no doubt of the superiority of skillfully made glass doublets and triplets. The idea is now, in fact, abandoned; and the same remark is applicable to the attempts at constructing fluid lenses, and to the projects for giving to glass other than spherical surfaces—none of which have come into extensive use.

By the term *simple* microscope is meant one in which the object is viewed directly through a lens or combination of lenses, just as we have supposed an arrow or an insect to be viewed through a glass held in the hand. When, however, the magnifying power of the glass is considerable, in other words, when its focal length is very short, and its proper distance from its object of consequence equally short, it requires to be placed at that proper distance with great precision: it cannot, therefore, be held with sufficient accuracy and steadiness by the unassisted hand, but must be mounted in a frame having a rack or screw to move it towards or from another frame or stage which holds the object. It is then called a microscope, and it is furnished, according to circumstances, with lenses and mirrors to collect and reflect the light upon the object, and with other conveniences which will now be described.

One of the best forms of a stand for a simple microscope is shown in Fig. 10, where A is a brass pillar screwed to a tripod base; B is a broad stage for the objects, secured to the stem by screws, whose milled heads are at C. By means of the large milled head D, a triangular bar, having a rack, is elevated out of the stem A, carrying the lens-holder E, which has a horizontal movement in one direction, by means of a rack worked by the milled head F, and in the other direction by turning on a circular pin. A concave mirror G reflects the light upwards through the hole in the stage, and a lens may be attached to the stage for the purpose of throwing light on an opaque object, in the same way that the forceps H for holding such objects is attached. This microscope is peculiarly adapted, by its broad stage and its general steadiness, for dissecting; and it is rendered more convenient for this purpose by placing it between two inclined planes of mahogany, which support the arms and elevate the wrists to the level of the stage. This apparatus is called the dissecting rest. When dissecting is not a primary object, a joint may be made at the lower end of the stem A, to allow the whole to take an inclined position; and then the spring clips shown upon the stage are useful to retain the object in its place. Numerous convenient appendages may be made to accompany such microscopes, which it will be impossible to mention in detail; the most useful are Mr. Varley's capillary cages for containing animalculæ in water, and parts of aquatic plants; also his tubes for obtaining and separating such

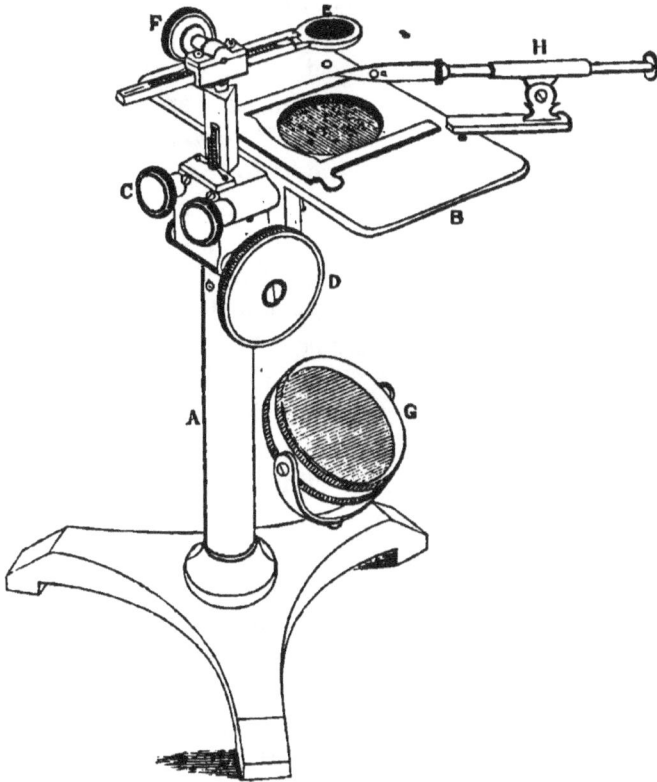

Fig. 10.

objects, and his phial and phial-holder for preserving and exhibiting small living specimens of the Chara, Nitella, and other similar plants, and observing their circulation. The phial-microscope affords facilities for observing the operations of minute vegetable and animal life, which will probably lead to the most interesting discoveries. The recent volumes of the Transactions of the Society of Arts contain an immense mass of information of this sort, and to these we refer the reader.

The mode of illuminating objects is one on which we must give some further information, for the manner in which an object is lighted is second in importance only to the excellence of the glass through which it is seen. In investigating any new or unknown specimen, it should be viewed in turns by every description of light, direct and oblique, as a transparent object and as an opaque object, with strong and with faint light, with large angular pencils and with small angular pencils thrown in all possible directions. Every change will probably develop some new fact in reference to the structure of the object, which should itself be varied in the mode of mounting in every possible way. It should be seen both wet and dry, and immersed in fluids of various qualities and densities, such as water, alcohol, oil, and Canada balsam, for instance, which last has a refractive power nearly equal to that of glass. If the object be delicate vegetable tissue, it will be in some respects rendered more visible by gentle heating or scorching by a clear fire placed between two plates of glass. In this way the spiral vessels of asparagus and other similar vegetables may be beautifully displayed. Dyeing the objects in tincture of iodine will in some cases answer this purpose better.

But the principal question in regard to illumination is the magnitude of the illuminating pencil, particularly in reference to transparent objects. Generally speaking the illuminating pencil should be as large as can be received by the lens, and no larger. Any light beyond this produces indistinctness and glare. The superfluous light from the mirror may be cut off by a screen having various-sized apertures placed below the stage; but the best mode of illumination is that proposed by Dr. Wollaston, and called the Wollaston condenser. A tube is placed below the stage of the instrument containing a lens A B (Fig. 11), which can be elevated or depressed within certain limits at pleasure; and at the lower end is a stop with a limited aperture C D. A plane mirror E F receives the rays of light L L from the sky or a white cloud, which last is the best source of light, and reflects them upwards through the aperture in C D, so that they are refracted, and form an image of the aperture at G, which is supposed to be nearly the place of the object. The object is sometimes best seen when the image of the aperture is also best seen; and sometimes it is best to elevate the summit G of the cone A B G above the object, and at others to depress it below: all which is done at pleasure by the power of moving the lens A B. If artifical light (as a lamp or candle) be employed, the flame must be placed in the principal focus of a large detached lens on a stand, so that the rays L L may fall in parallel lines on the mirror, or as they would fall from the cloud. This will be found an advantage, not only when the Wollaston condenser is employed, but also when the mirror and diaphragm are used. A good mode of imitating artificially the light of a white cloud opposite the sun has been proposed by Mr. Varley; he covers the surface of the mirror under the stage with carbonate of soda or any similar material, and then concentrates the sun's light upon its surface by a large condensing lens. The intense white light diffused from the surface of the soda forms an excellent substitute for the white cloud, which, when opposite the sun, and of considerable size, is the best daylight, as the pure sky opposite to the sun is the worst.

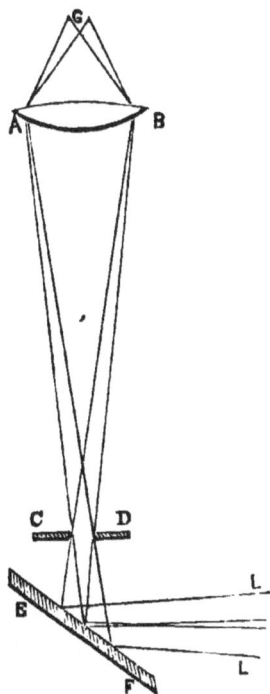

Fig. 11.

The Compound Microscope may, as before stated, consist of only two lenses, while a simple microscope has been shown to contain sometimes three. In the triplet for the simple microscope, however, it was explained that the effect of the two first lenses was to do what might have been accomplished, though not so well, by one; and the third merely effected certain modifications in the light before it entered the eye. But in the compound microscope the two lenses have totally different functions; the first receives the rays from the object, and, bringing them to new

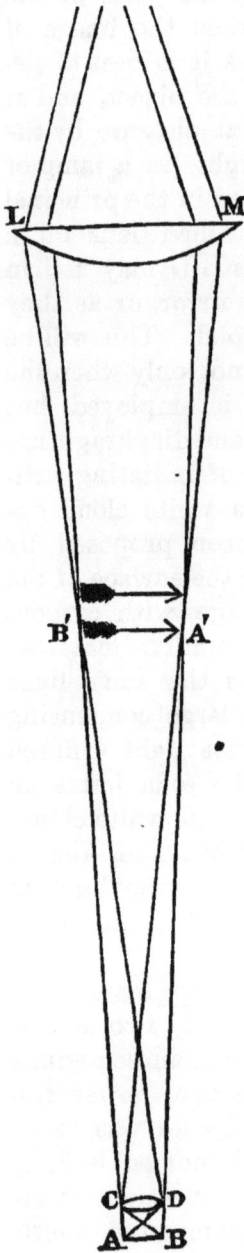

foci, forms an image, which the second lens treats as an original object, and magnifies it just as the single microscope magnified the object itself.

The annexed figure (12) shows the course of the rays through a compound microscope of two lenses. The rays proceeding from the object A B are so acted upon by the lens C D, near it, and thence called the object glass, that they are converged to foci in A′ B′, where they form an enlarged image of the object, as would be evident if a piece of oiled paper or ground glass were placed there to receive them. They are not so intercepted, and therefore the image is not rendered visible at that place; but their further progress is similar to what it would have been had they really proceeded from an object at A′ B′. They are at length received by the eye-lens L M, which acts upon them as the simple microscope has been described to act on the light proceeding from its objects. They are bent so that they may enter the eye at E in parallel lines, or as nearly so as is requisite for distinct vision. When we say that the rays enter the eye in nearly parallel lines, we mean only those which proceed from one point of the original object. Thus the two parallel rays M E have proceeded from and are part of the cone of rays C A D, emanating from the point A of the arrow; but they do not form two pictures in the eye, because any number of parallel rays which the pupil can receive will be converged to a point by the eye, and will convey the impression· of one point to the mind. In like manner the

Fig. 12.

rays L E are part of the cone of rays emanating from B, and the angle L E M is that under which the eye will see the magnified image of the arrow, which is evidently many times greater than the arrow could be made to occupy in the naked eye at any distance within the limits of distinct vision. The magnifying power depends on two circumstances: first, on the ratio between the anterior distance A O or B D and the posterior focal length C B' or D A'; and secondly, on the power of the eye-lens L M. The first ratio is the same as that between the object A B and the image A' B'; this and the focal length or power of the eye lens are both easily obtained, and their product is the power of the compound instrument.

Since the power depends on the ratio between the anterior and posterior foci of the object-glass, it is evident that by increasing that ratio any power may be obtained, the same eye-glass being used; or having determined the first, any further power may be obtained by increasing that of the eye-glass; and thus, by a pre-arrangement of the relative proportions in which the magnifying power shall be divided between the object-glass and the eye-glass, almost any given distance (within certain limits) between the first and its object may be secured. This is one valuable peculiarity of the compound instrument; and another is the large field, or large angle of view, which may be obtained, every part of which will be nearly equally good; whereas with the best simple microscopes the field is small, and is good only in the centre. The field of the compound instrument is further increased by using two glasses at the eye-end; the first being called, from its purpose, the field-glass, and the two constituting what is called the eye-piece. This will be more particularly explained in the figure of the achromatic compound microscope presently given.

For upwards of a century the compound microscope, notwithstanding the advantages above mentioned, was a comparatively feeble and inefficient instrument, owing to the distance which the light had to traverse, and the consequent increase of the chromatic and spherical aberrations. To explain this we have drawn in Fig. 12 a second image near A' B', the fact being that the object-glass would not form one image, as has been supposed, but an infinite number of variously-colored and vari-

ous-sized images, occupying the space between the two dotted
arrows. Those nearest the object-glass would be red, and those
nearest the eye-glass would be blue. The effect of this is to
produce so much confusion, that the instrument was reduced
to a mere toy, although these errors were diminished to the
utmost possible extent by limiting the aperture of the object-
glass, and thus restricting the angle of the pencil of light from
each point of the object. But this involved the defects, already
explained, of making the picture obscure, so that on the whole
the best compound instruments were inferior to the simple
microscopes of a single lens, with which, indeed, all the important observations of the last century were made.

Even after the improvement of the simple microscope by the
use of doublets and triplets, the long course of the rays, and
the large angular pencil required in the compound instrument,
deterred the most sanguine from anticipating the period when
they should be conducted through such a path free both from
spherical and chromatic errors. Within twenty years of the
present period, philosophers of no less eminence than M. Blot
and Dr. Wollaston predicted that the compound would never
rival the simple microscope, and that the idea of achromatizing
its object-glass was hopeless. Nor can these opinions be won-
dered at when we consider how many years the achromatic
telescope had existed without an attempt to apply its principles
to the compound microscope. When we consider the smallness
of the pencil required by the telescope, and the enormous increase of difficulty attending every enlargement of the pencil—
when we consider further that these difficulties had to be con-
tended with and removed by operations on portions of glass so
small that they are themselves almost microscopic objects, we
shall not be surprised that even a cautious philosopher and most
able manipulator like Dr. Wollaston should prescribe limits to
improvement.

Fortunately for science, and especially for the departments
of animal and vegetable physiology, these predictions have
been shown to be unfounded. The last fifteen years have suf-
ficed to elevate the compound microscope from the condition
we have described to that of being the most important instrument ever bestowed by art upon the investigator of nature. It

now holds a very high rank among philosophical implements, while the transcendant beauties of form, color and organization, which it reveals to us in the minute works of nature, render it subservient to the most delightful and instructive pursuits. To these claims on our attention, it appears likely to add a third of still higher importance. The microscopic examination of the blood and other human organic matter will in all probability afford more satisfactory and conclusive evidence regarding the nature and seat of disease than any hitherto appealed to, and will of consequence lead to similar certainty in the choice and application of remedies.

We have thought it necessary to state thus at large the claims of the modern achromatic microscope upon the attention of the reader, as a justification of the length at which we shall give its recent history and explain its construction; and we are further induced to this course by the consideration that the subject is entirely new ground, and that there are at this time not more than two or three makers of achromatic microscopes in England.

Soon after the year 1820 a series of experiments was begun in France by M. Selligues, which were followed up by Frauenhofer at Munich, by Amici at Modena, by M. Chevalier at Paris, and by the late Mr. Tulley in London. In 1824 the last-named excellent artist, without knowing what had been done on the Continent, made the attempt to construct an achromatic object-glass for a compound microscope, and produced one of nine-tenths of an inch focal length, composed of three lenses, and transmitting a pencil of eighteen degrees. This was the first that had been made in England; and it is due to Mr. Tulley to say, that as regards accurate correction throughout the field, that glass has not been excelled by any subsequent combination of three lenses. Such an angular pencil, and such a focal length, would bear an eye-piece adapted to produce a gross magnifying power of one hundred and twenty. Mr. Tulley afterwards made a combination to be placed in front of the first mentioned, which increased the angle of the transmitted pencil to thirty-eight degrees, and bore a power of three hundred.

While these practical investigations were in progress, the

subject of achromatism engaged the attention of some of the most profound mathematicians in England. Sir John Herschel, Professor Airy, Professor Barlow, Mr. Coddington, and others, contributed largely to the theoretical examination of the subject; and though the results of their labors were not immediately applicable to the microscope, they essentially promoted its improvement.

For some time prior to 1829 the subject had occupied the mind of a gentleman, who, not entirely practical, like the first, nor purely mathematical, like the last-mentioned class of inquirers, was led to the discovery of certain properties in achromatic combinations which had been before unobserved. These were afterwards experimentally verified; and in the year 1829 a paper on the subject, by the discoverer, Mr. Joseph Jackson Lister, was read and published by the Royal Society. The principles and results thus obtained enabled Mr. Lister to form a combination of lenses which transmitted a pencil of fifty degrees, with a large field correct in every part; as this paper was the foundation of the recent improvements in acromatic microscopes, and as its results are indispensable to all who would make or understand the instrument, we shall give the more important parts of it in detail, and in Mr. Lister's own words.

"I would premise that the plano-concave form for the correcting flint lens has in that quality a strong recommendation, particularly as it obviates the danger of error which otherwise exists in centering the two curves, and thereby admits of correct workmanship for a shorter focus. To cement together also the two surfaces of the glass diminishes by very nearly half the loss of light from reflection, which is considerable at the numerous surfaces of a combination. I have thought the clearness of the field and brightness of the picture evidently increased by doing this; it prevents any dewiness or vegetation from forming on the inner surfaces; and I see no disadvantage to be anticipated from it if they are of identical curves, and pressed closely together, and the cementing medium permanently homogeneous.

"These two conditions then, that the flint lens shall be plano-concave, and that it shall be joined by some cement to the convex, seem desirable to be taken as a basis for the microscopic object-glass, provided they can be reconciled with the destruction of the spherical and chromatic aberrations of a large pencil.

"Now in every such glass that has been tried by me which has had its correcting lens of either Swiss or English glass, with a double convex of plate, and has been made achromatic by the form given to the outer curve of the convex, the proportion has been such between the refractive and dispersive powers of its lenses, that its figure has been correct for rays issuing from some point in its axis not far from its principal focus on its plane side, and either tending to a conjugate focus within the tube of a microscope, or emerging nearly parallel.

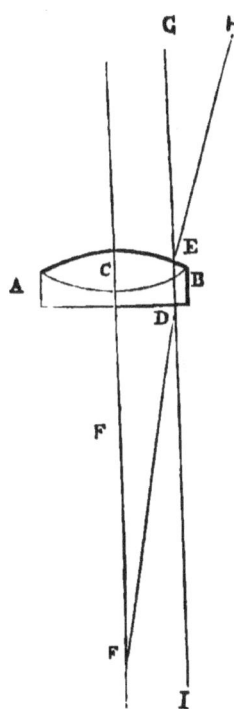

"Let A B (Fig. 13) be supposed such an object-glass, and let it be roughly considered as a plano-convex lens, with a curve A C B running through it, at which the spherical and chromatic errors are corrected which are generated at the two outer surfaces; and let the glass be thus free from aberration for rays F D E G issuing from the radiant point F, H E being a perpendicular to the convex surface, and I D to the plane one. Under these circumstances, the angle of emergence G E H much exceeds that of incidence F D I, being probably nearly three times as great.

Fig. 13.

"If the radiant is now made to approach the glass, so that the course of the ray F D E G shall be more divergent from the axis, as the angles of incidence and emergence become more nearly equal to each other, the spherical aberration produced by the two will be found to bear a less proportion to the opposing error of the single correcting curve A C B; for such a focus therefore the rays will be over-corrected.

"But if F still approaches the glass, the angle of incidence continues to increase with the increasing divergence of the ray, till it will exceed that of emergence, which has in the meanwhile been diminishing, and at length the spherical error produced by them will recover its original proportion to the opposite error of the curve of correction. When F has reached this point F″ (at which the angle of incidence does not exceed that of emergence so much as it had at first come short of it), the rays again pass the glass free from spherical aberration.

"If F be carried from hence towards the glass, or outwards from its original place, the angle of incidence in the former case, or of emergence in the latter, becomes disproportionately effective, and either way the aberration exceeds the correction.

"These facts have been established by careful experiment: they accord with every appearance in such combinations of the plano-convex glasses as have come under my notice, and may, I believe, be extended to this rule, that in general an achromatic object-glass, of which the inner surfaces are in contact, or nearly so, will have on one side of it two foci in its axis, for the rays proceeding from which it will be truly corrected at a moderate aperture; that for the space between these two points its spherical aberration will be over-corrected, and beyond them either way under-corrected.

"The longer aplanatic focus may be found, when one of the plano-convex object-glasses is placed in a microscope, by shortening the tube, if the glass shows over-correction; if under-correction, by lengthening it, or by bringing the rays together, should they be parallel or divergent, by a very small good telescope. The shorter focus is got at by sliding the glass before another of sufficient length and large aperture that is finely corrected, and bringing it forwards till it gives the reflection of a bright point from a globule of quicksilver, sharp and free from mist, when the distance can be taken between the glass and the object.

"The longer focus is the place at which to ascertain the utmost aperture that may be given to the glass, and where, in the absence of spherical error, its exact state of correction as to color is seen most distinctly.

"The correction of the chromatic aberration, like that of the spherical, tends to excess in the marginal rays; so that if a glass which is achromatic, with a moderate aperture, has it cell opened wider, the circle of rays thus added to the penci will be rather over-corrected as to color.

"The same tendency to over-correction is produced, if, without varying the aperture, the divergence of the incident rays is much augmented, as in an object-glass placed in front of another; but generally in this position a part only of its aperture comes into use; so that the two properties mentioned neutralize each other, and its chromatic state remains unaltered. If, for example, the outstanding colors were observed at the longer focus to be green and claret, which show that the nearest practicable approach is made to the union of the spectrum, they usually continue nearly the same for the whole space between the foci, and for some distance beyond them either way.

"The places of these two foci and their proportions to each other depend on a variety of circumstances. In several object-glasses that I have had made for trial, plano-convex, with their inner surfaces cemented, their diameters the radius of the flint lens, and their color pretty well corrected, those composed of dense flint and light plate have had the rays from the longer focus emerging nearly parallel; and this focus has been not quite three times the distance of the shorter from the glass: with English flint the rays have had more convergence, and the shorter focus has borne a rather less proportion to the longer.

"If the surfaces are not cemented, a striking effect is produced by minute differences in their curves. It may give some idea of this, that in a glass of which nearly the whole disk was covered with color from contact of the lenses, the addition of a film of varnish, so thin that this color was not destroyed by it, caused a sensible change in the spherical correction.

"I have found that whatever extended the longer aplanatic focus, and increased the convergence of its rays, diminished the relative length of the shorter. Thus by turning to the concave lens the flatter instead of the deeper side of a convex lens, whose radii were to each other as 31 to 35, the pencil of the longer aplanatic focus, from being greatly divergent, was brought to converge at a very small distance behind the glass; and the length of the shorter focus, which had been one-half that of the longer, became but one-sixth of it.

"The direction of the aplanatic pencils appears to be scarcely affected by the differences in the thickness of glasses, if their state as to color is the same.

"One other property of the double object-glass remains to be mentioned, which is, that when the longer aplanatic focus is used, the marginal rays of a pencil not coincident with the axis of the glass are distorted, so that a coma is thrown outwards; while the contrary effect of a coma directed towards the centre of the field is produced by the rays from the shorter focus. These peculiarities of the coma seem inseparable attendants on the two foci, and are as conspicuous in the achromatic meniscus as in the plano-convex object-glass.

"Of several purposes to which the particulars just given seem applicable, I must at present confine myself to the most obvious one. They furnish the means of destroying with the utmost ease both aberrations in a large focal pencil, and of thus surmounting what has hitherto been the chief obstacle to the perfection of the microscope. And when it is considered that the curves of its diminutive object-glasses have required to be at least as exactly proportioned as those of a large telescope to give the image of a bright point equally sharp and colorless, and that any change made to correct one aberration was liable to disturb the other, some idea may be formed of what the amount of that obstacle must have been. It will, however, be evident that if any object-glass is but made achromatic, with its lenses truly worked and cemented, so that their axes coincide, it may with certainty be connected with another possessing the same requisites and of suitable focus, so that the combination shall be free from spherical error also in the centre of its field. For this the rays have only to be received by the front glass B (Fig. 14) from its shorter aplanatic focus F''', and transmitted in the direction of the longer correct pencil F A of the other glass A. It is desirable that the latter pencil should neither converge to a very short focus nor be more than

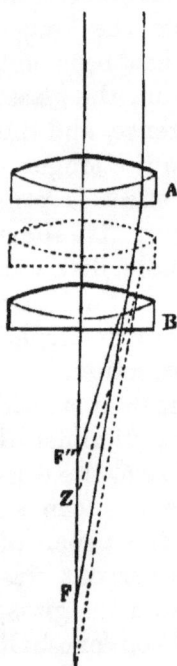

Fig. 14.

very slightly if at all divergent; and a little attention at first to the kind of glass used will keep it within this range, the denser flint being suited to the glasses of shorter focus and larger angle of aperture.

"The adjustment of the microscope is then perfected, if necessary, by slightly varying the distance between the object-glasses; and after that is done, the length of the tube which carries the eye-pieces may be altered greatly without disturbing the correction, opposite errors which balance each other being produced by the change.

"If the two glasses which in the diagram are drawn at some distance apart are brought nearer together (if the place of A, for instance, is carried to the dotted figure), the rays transmitted by B in the direction of the longer aplanatic pencil of A will plainly be derived from some point Z more distant than F‴, and lying between the aplanatic foci of B; therefore (according to what has been stated) this glass, and consequently the combination, will then be spherically over-corrected. If, on the other hand, the distance between A and B is increased, the opposite effects are of course produced.

"In combining several glasses together it is often convenient to transmit an under-corrected pencil from the front glass, and to counteract its error by over-correction in the middle one.

"Slight errors in color may in the same manner be destroyed by opposite ones; and on the principles described we not only acquire fine correction for the central ray, but by the opposite effects at the two foci on the transverse pencil, all coma can be destroyed, and the whole field rendered beautifully flat and distinct."

Mr. Lister's paper enters into further particulars, which are not essential to the comprehension of the subject. It is sufficient to say that his investigations and results proved to be of the highest value to the practical optician, and the progress of improvement was in consequence extremely rapid. The new principles were applied and exhibited by Mr. Hugh Powell and Mr. Andrew Ross with a degree of success which had never been anticipated; so perfect indeed were the corrections given to the achromatic object-glass—so completely were the errors of sphericity and dispersion balanced or destroyed—that the circumstance of covering the object with a plate of the thinnest glass or talc disturbed the corrections, if they had been adapted to an uncovered object, and rendered an object-glass which was perfect under one condition sensibly defective under the other.

This defect, if that should be called a defect which arose out of improvement, was first discovered by Mr. Ross, who immediately suggested the means of correcting it, and presented to the Society of Arts, in 1837, a paper on the subject, which was published in the 51st volume of their Transactions, and which, as it is, like Mr. Lister's essential to a full understanding of the ultimate refinements of the instrument, we shall extract nearly in full:

"In the course of a practical investigation (says Mr. Ross) with the view of constructing a combination of lenses for the object-glass of a compound microscope, which should be free from the effects of aberration, both for central and oblique pencils of great angle, I combined the condition of the greatest possible distance between the object and object-glass; for in object-glasses of short focal length their closeness to the object has been an obstacle in many cases to the use of high magnifying powers, and is a constant source of inconvenience.

"In the improved combination, the diameter is only sufficient to admit the proper pencil; the convex lenses are wrought to an edge, and the concave have only sufficient thickness to support their figure; consequently the combination is the thinnest possible, and it follows that there will be the greatest distance between the object and the object-glass. The focal length is one-eighth of an inch, having an angular aperture of 60°, with a distance of 1-25th of an inch, and a magnifying power of 970 times linear, with perfect definition on the most difficult Podura scales. I have made object-glasses 1-16th of an inch focal length; but as the angular aperture cannot be advantageously increased, if the greatest distance between the object and object-glass is preserved, their use will be very limited.

"The quality of the definition produced by an achromatic compound microscope will depend upon the accuracy with which the aberrations, both chromatic and spherical, are balanced, together with the general perfection of the workmanship. Now, in Wollaston's doublets, and Holland's triplets, there are no means of producing a balance of the aberrations, as they are composed of convex lenses only; therefore the best that can be done is to make the aberrations a minimum; the remaining positive aberration in these forms produces its peculiar effect upon objects (particularly the detail of the thin transparent class), which may lead to misapprehension of their true structure; but with the achromatic object-glass, where the aberrations are correctly balanced, the most minute parts of an object are accurately displayed, so that a satisfactory judgment of their character may be formed.

"It will be seen by Fig. 15, that when a certain angular pencil A O A' proceeds from the object O, and is incident on the plane side of the first lens, if the combination is removed from

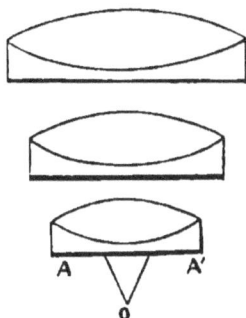 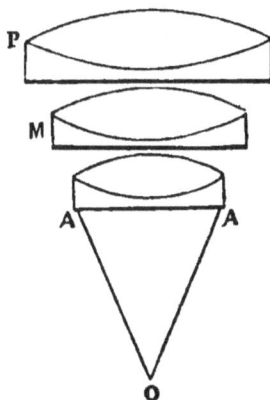

Fig. 15.   Fig. 16.

the object, as in Fig. 16, the extreme rays of the pencil impinge on the more marginal parts of the glass, and as the refractions are greater here, the aberrations will be greater also. Now, if two compound object-glasses have their aberrations balanced, one being situated as in Fig. 15, and the other as in Fig. 16, and the same disturbing power applied to both, that in which the angles of incidence and the aberrations are small will not be so much disturbed as where the angles are great, and where consequently the aberrations increase rapidly.

"When an object-glass has its aberrations balanced for viewing an opaque object, and it is required to examine that object
by transmitted light, the correction will remain; but if it is
necessary to immerse the object in a fluid, or to cover it with
glass or talc, an aberration will arise from these circumstances,
which will disturb the previous correction, and consequently
deteriorate the definition; and this effect will be more obvious
with the increase of the distance between the object and the
object-glass.

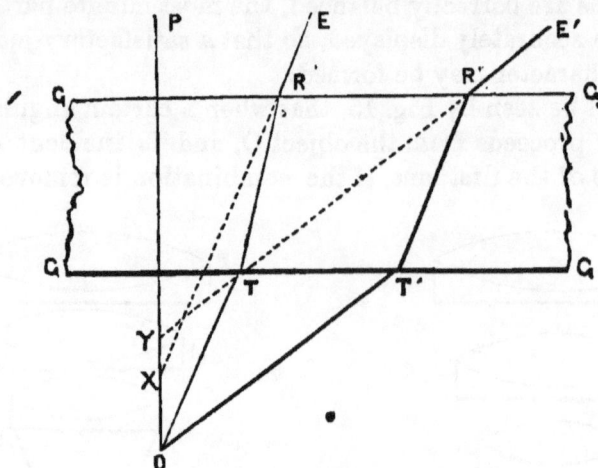

Fig. 17.

"The aberration produced with diverging rays by a piece of
flat and parallel glass, such as would be used for covering an
object, is represented at Fig. 17, where G G G G is the refracting medium, or piece of glass covering the object O; O P, the
axis of the pencil, perpendicular to the flat surfaces; O T, a
ray near the axis; and O T', the extreme ray of the pencil incident on the under surface of the glass; then T R, T' R', will be
the directions of the rays in the medium, and R E, R' E', those
of the emergent rays. Now if the course of these rays is con-
tinued, as by the dotted lines, they will be found to intersect
the axis at different distances, X and Y, from the surface of the
glass; and the distance X Y is the aberration produced by the
medium which, as before stated, interferes with the previously
balanced aberrations of the several lenses composing the object-

glass. There are many cases of this, but the one here selected serves best to illustrate the principle. I need not encumber the description with the theoretical determination of this quantity, as it varies with exceedingly minute circumstances which we cannot accurately control; such as the distance of the object from the under side of the glass, and the slightest difference in the thickness of the glass itself; and if these data could be readily obtained, the knowledge would be of no utility in making the correction, that being wholly of a practical nature.

"If an object-glass is constructed as represented in Fig. 16, where the posterior combination P and the middle M have together an excess of negative aberration, and if this be corrected by the anterior combination A, having an excess of positive aberration, then this latter combination can be made to act more or less powerfully upon P and M, by making it approach to or recede from them; for when the three are in close contact, the distance of the object from the object-glass is greatest; and consequently the rays from the object are diverging from a point at a greater distance than when the combinations are separated; and as a lens bends the rays more, or acts with greater effect, the more distant the object is from which the rays diverge, the effect of the anterior combination A upon the other two, P and M, will vary with its distance from thence. When therefore the correction of the whole is effected for an opaque object with a certain distance between the anterior and middle combination, if they are then put in contact, the distance between the object and object-glass will be increased; consequently the anterior combination will act more powerfully, and the whole will have an excess of positive aberration. Now the effect of the aberration produced by a piece of flat and parallel glass being of the negative character, it is obvious that the above considerations suggest the means of correction by moving the lenses nearer together, till the positive aberration thereby produced balances the negative aberration caused by the medium.

"The preceding refers only to the spherical aberration, but the effect of the chromatic is also seen when an object is covered with a piece of glass; for, in the course of my experiments, I observed that it produced a chromatic thickening of the out-

line of the Podura and other delicate scales; and if diverging
rays near the axis and at the margin are projected through a
piece of flat parallel glass, with the various indices of refraction
for the different colors, it will be seen that each ray will emerge
separated into a beam consisting of the component colors of
the ray, and that each beam is widely different in form. This
difference, being magnified by the power of the microscope,
readily accounts for the chromatic thickening of the outline
just mentioned. Therefore to obtain the finest definition of
extremely delicate and minute objects, they should be viewed
without a covering; if it be desirable to immerse them in a fluid,
they should be covered with the thinnest possible film of talc,
as, from the character of the chromatic aberration, it will be
seen that varying the distances of the combinations will not
sensibly affect the correction; though object-lenses may be
made to include a given fluid or solid medium in their correction for color.

"The mechanism for applying these principles to the correction of an object-glass under the various circumstances, is
represented in Fig. 18, where the anterior lens is set in the end

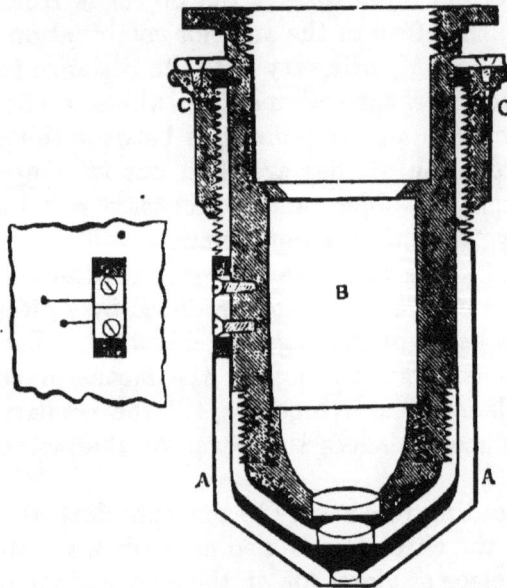

Fig. 18.

of a tube A A, which slides on the cylinder B containing the remainder of the combination; the tube A A, holding the lens nearest the object, may then be moved upon the cylinder B, for the purpose of varying the distance according to the thickness of the glass covering the object, by turning the screwed ring C C, or more simply by sliding the one on the other, and clamping them together when adjusted. An aperture is made in the tube A, within which is seen a mark engraved on the cylinder, and on the edge of which are two marks, a longer and a shorter, engraved upon the tube. When the mark on the cylinder coincides with the longer mark on the tube, the adjustment is perfect for an uncovered object; and when the coincidence is with the short mark, the proper distance is obtained to balance the aberrations produced by glass one-hundredth of an inch thick, and such glass can be readily supplied.

"It is hardly necessary to observe, that the necessity for this correction is wholly independent of any particular construction of the object-glass; as in all cases where the object-glass is corrected for an object uncovered, any covering of glass will create a different value of aberration to the first lens, which previously balanced the aberration resulting from the rest of the lenses; and as this disturbance is effected at the first refraction, it is independent of the other part of the combination. The visibility of the effect depends on the distance of the object from the object-glass, the angle of the pencil transmitted, the focal length of the combination, the thickness of the glass covering the object, and the general perfection of the corrections for chromatism and the oblique pencils.

"With this adjusting object-glass, therefore, we can have the requisites of the greatest possible distance between the object and object-glass, an intense and sharply defined image throughout the field from the large pencil transmitted, and the accurate correction of the aberrations; also, by the adjustment, the means of preserving that correction under all the varied circumstances in which it may be necessary to place an object for the purpose of observation."

In the annexed engraving, Fig. 19, we have shown the triple achromatic object-glass in connection with the eye-piece consisting of the field-glass F F, and the eye-glass E E, forming together the modern achromatic microscope. The course of the light is shown by drawing three rays from the centre and three from each end of the object O. These rays would, if left to themselves, form an image of the object at A A, but being bent and converged by the field-glass F F, they form the image at B B, where a stop is placed to intercept all light except what is required for the formation of the image. From B B therefore the rays proceed to the eye-glass exactly as has been described in reference to the simple microscope and to the compound of two glasses.

If we stopped here we should convey a very imperfect idea of the beautiful series of corrections effected by the eye-piece, and which were first pointed out in detail in a paper on the subject published by Mr. Varley in the 51st volume of the Transactions of the Society of Arts. The eye-piece in question was invented by Huyghens for telescopes, with no other view than that of diminishing the spherical aberration by producing the refractions at two glasses instead of one, and of increasing the field of view. It was reserved for Boscovich to point out that Huyghens had by this arrangement accidentally corrected a great part of the chromatic aberration, and this subject is further investigated with much skill

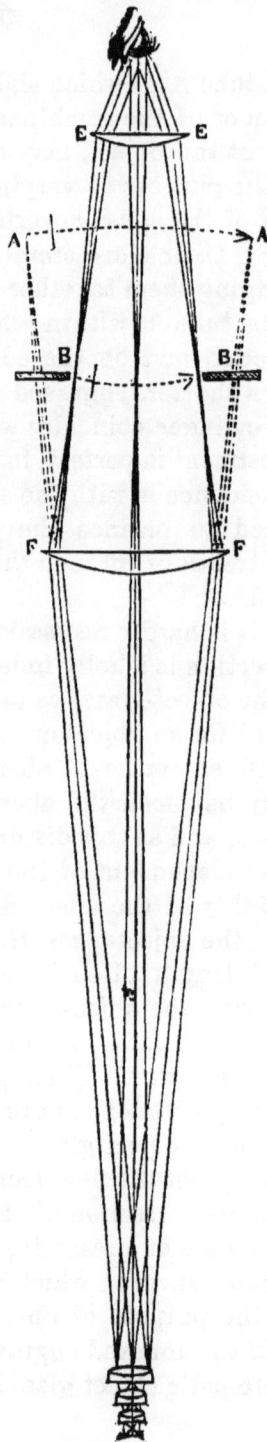

Fig. 19.

in two papers by Professor Airy in the *Cambridge Philosophical Transactions*, to which we refer the mathematical reader. These investigations apply chiefly to the telescope, where the small pencils of light and great distance of the object exclude considerations which become important in the microscope, and which are well pointed out in Mr. Varley's paper before mentioned.

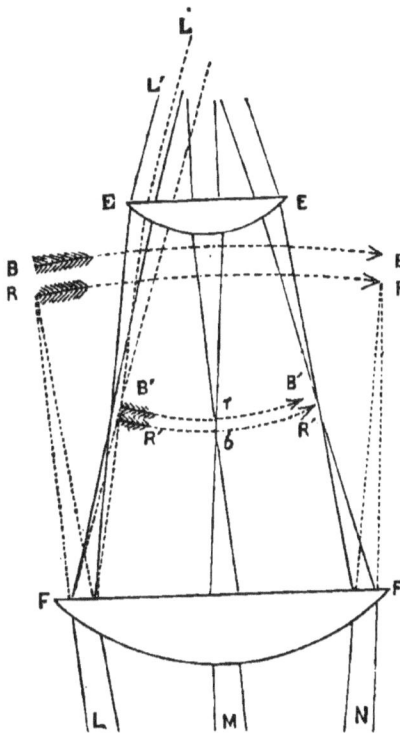

Fig. 20.

Let Fig. 20 represent the Huyghenean eye-piece of a microscope; F F and E E being the field-glass and eye-glass, and L M N the two extreme rays of each of the three pencils, emanating from the centre and ends of the object, of which, but for the field-glass, a series of colored images would be formed from R R to B B; those near R R being red, those near B B blue, and the intermediate ones green, yellow, and so on, corresponding with the colors of the prismatic spectrum. This order of colors, it will be observed, is the reverse of that described in treating of the common compound microscope (Fig. 12), in which the single object-glass projected the red image beyond the blue. The effect just described, of projecting the blue image beyond the red, is purposely produced for reasons presently to be given, and is called over-correcting the object-glass as to color. It is to be observed also that the images B B and R R are curved in the wrong direction to be distinctly seen by a convex eye-lens, and this is a further defect of the compound microscope of two lenses. But the field-glass, at the same time that it bends the rays and converges them to foci at B' B' and R' R', also reverses the curvature of the images as there shown, and gives them the form best adapted for distinct vision by the eye-glass E E. The field-glass has at the same time brought the blue and red images closer together, so that they are adapted to pass uncolored through the eye-glass. To render this important point more intelligible, let it be supposed that the object-glass had not been over-corrected, that it had been perfectly achromatic; the rays would then have become colored as soon as they had passed the field-glass; the blue rays, to take the central pencil, for example, would converge at *b* and the red rays at *r*, which is just the reverse of what the eye-lens requires; for as its blue focus is also shorter than its red, it would demand rather that the blue image should be at *r* and the red at *b*. This effect we have shown to be produced by the over-correction of the object-glass, which protrudes the blue foci B B as much beyond the red foci R R as the sum of the distances between the red and blue foci of the field-lens and eye-lens; so that the separation B R is exactly taken up in passing through those two lenses, and the whole of the colors coincide as to focal distance as soon as the rays have passed the eye-lens. But while they coincide as to distance, they differ in another respect; the blue images are rendered smaller than the red by the superior refractive power of the field-glass upon the blue rays. In tracing the pencil L, for instance, it will be noticed that after passing the field-glass, two sets of lines are drawn, one whole, and one dotted, the former representing the red, and the latter the blue rays. This is the accidental effect in the Huyghenean eye-piece pointed out by Boscovich. This separation into colors at the field-glass is like the over-correction of the object-glass; it leads to a subsequent complete correction. For if the differently colored rays were kept together till they reached the eye-glass, they would then become colored, and present colored images to the eye; but fortunately, and most beautifully, the separation effected by the field-glass causes the blue rays to fall so much nearer the centre of the eye-glass, where, owing to the spherical figure, the refractive power is less than at the margin, that the spherical error of the eye-lens constitutes a nearly perfect balance to the chromatic dispersion of the field-lens, and the red and blue rays L' and L" emerge sensibly parallel, presenting, in consequence, the perfect definition of a single point to the eye. The same reasoning is true of the intermediate colors and of the other pencils.

From what has been stated it is obvious that we mean by an achromatic object-glass one in which the usual order of dispersion is so far reversed that the light, after undergoing the singularly beautiful series of changes effected by the eye-piece, shall come uncolored to the eye. We can give no specific rules for producing these results. Close study of the formulæ for achromatism given by the celebrated mathematicians we have quoted will do much, but the principles must be brought to the test of repeated experiment. Nor will the experiments be worth anything, unless the curves be most accurately measured and worked, and the lenses centered and adjusted with a degree of precision which, to those who are familiar only with telescopes, will be quite unprecedented.

The Huyghenean eye-piece which we have described is the best for merely optical purposes, but when it is required to measure the magnified image, we use the eye-piece invented by Mr. Ramsden, and called, from its purpose, the micrometer eye-piece. When it is stated that we sometimes require to measure portions of animal or vegetable matter a hundred times smaller than any divisions that can be artificially made on any measuring instrument, the advantage of applying the scale to the magnified image will be obvious, as compared with the application of engraved or mechanical micrometers to the stage of the instrument.

The arrangement is shown in Fig. 21, where E E and F F are the eye and field glass, the latter having now its plane face towards the object. The rays from the object are here made to converge at A A, immediately in front of the field-glass, and here also is placed a plane glass on which are engraved divisions of a hundredth of an inch or less. The markings of these divisions come into focus therefore at the same time as the image of the object, and both are distinctly seen together. Thus the measure of the magnified image is given by mere inspection, and the value of such measures in reference to the real object may be obtained thus, which, when once obtained, is constant for the same object-glass. Place on the stage of the instrument a divided scale the value of which is known, and viewing this scale as the microscopic object, observe how many of the divisions on the scale attached to the eye-piece correspond with one of those in the magnified image. If, for instance, ten of those in the eye-piece correspond with one of those in the image, and if the divisions are known to be equal, then the image is ten times larger than the object, and the dimensions of the object are ten times less than indicated by the micrometer. If the divisions on the micrometer and on the magnified scale were not equal, it becomes a mere rule-of-three sum, but in general this trouble is taken by the maker of the instrument, who furnishes a table showing the value of each division of the micrometer for every object-glass with which it may be used.

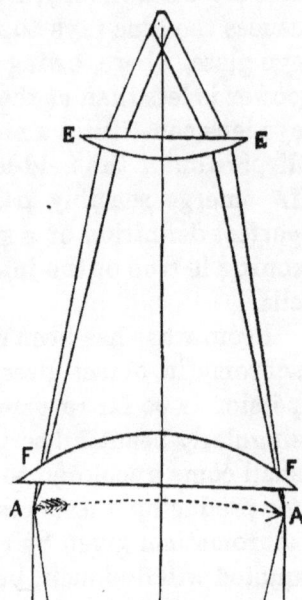

Fig. 21.

While on the subject of measuring it may be well to explain the mode of ascertaining the magnifying power of the compound microscope, which is generally taken on the assumption before mentioned, that the naked eye sees most distinctly at the distance of ten inches.

Place on the stage of the instrument, as before, a known divided scale, and when it is distinctly seen, hold a rule at ten inches distance from the disengaged eye, so that it may be seen by that eye, overlapping or lying by side of the magnified picture of the other scale. Then move the rule till one or more of its known divisions correspond with a number of those in the magnified scale, and a comparison of the two gives the magnifying power.

Having now explained the optical principles of the achromatic compound microscope, it remains only to describe the mechanical arrangements for giving those principles their full effect. The mechanism of a microscope is of much more importance than might be imagined by those who have not studied the subject. In the first place, steadiness, or freedom from vibration, and most particularly freedom from any vibrations which are not equally communicated to the object under examination, and to the lenses by which it is viewed, is a point of the utmost consequence. When, for instance, the body containing the lenses is screwed by its lower extremity to a horizontal arm, we have one of the most vibratory forms conceivable; it is precisely the form of the inverted pendulum, which is expressly contrived to indicate otherwise insensible vibrations. The tremor necessarily attendant on such an arrangement is magnified by the whole power of the instrument; and as the object on the stage partakes of this tremor in a comparatively insensible degree, the image is seen to oscillate so rapidly, as in some cases to be wholly undistinguishable. Such microscopes cannot possibly be used with high powers in ordinary houses abutting on any paved streets through which carriages are passing, nor indeed are they adapted to be used in houses in which the ordinary internal sources of shaking exist.

One of the best modes of mounting a compound microscope is shown in the annexed view (Fig. 22), which, though too minute to exhibit all the details, will serve to explain the chief features of the arrangement.

A massy pillar A is screwed into a solid tripod B, and is surmounted by a strong joint at C, on which the whole instrument turns, so as to enable it to take a perfectly horizontal or vertical position, or any intermediate angle, such, for instance, as that shown in the engraving.

This movable portion of the instrument consists of one solid casting D E F G; from F to G being a thick pierced plate carrying the stage and its appendages. The compound body H is attached to the bar D E, and moves up and down upon it by a rack and pinion worked by either of the milled heads K. The piece D E F G is attached to the pillar by the joint C, which being the source of the required movement in the instrument, is obviously its weakest part, and about which no doubt considerable vibration takes place. But inasmuch as the piece D E F G of necessity transmits such vibrations equally to the body of the microscope and to the objects on the stage, they hold always the same relative position, and no *visible* vibration is caused, how much soever may really exist. To the under side of the stage is attached a circular stem L, on which slides the mirror M, plane on one side and concave on the other, to reflect the light through the aperture in the stage. Beneath the stage is a circular revolving plate containing three apertures of various sizes, to limit the angle of the pencil of light which shall be allowed to fall on the object under examination. Besides these conveniences the stage has a double movement produced by two racks at right angles to each other, and worked by milled heads beneath. It has also the usual appendages of forceps to hold minute objects, and a lens to condense the light upon them, all of which are well understood, and if not, will be rendered more intelligible by a few minutes' examination of a microscope than by the most lengthened description. One other point remains to be noticed. The movement produced by the milled head K is not sufficiently delicate to adjust the focus of very powerful lenses, nor indeed is any rack movement. Only the finest screws are adapted to this purpose; and even these are improved by means for reducing the rapidity of the screw's movement. For this purpose the lower end of the compound body H, which carries the object-glass, consists of a piece of smaller tube sliding in parallel guides in the main body, and kept constantly pressed upwards by a spiral spring but it can be drawn downward by a lever crossing the body, and acted on by an extremely fine screw

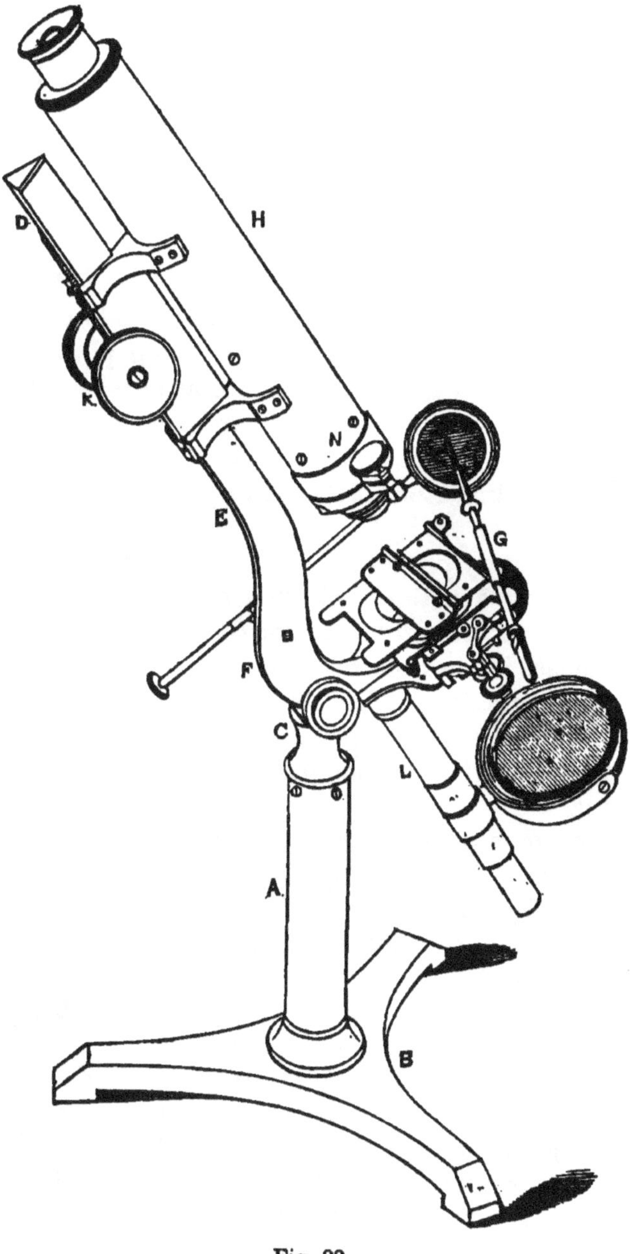

Fig. 22.

whose milled head is seen at N, and the fineness of which
is tripled by means of the lever through which it acts on the
object-glass. The instrument is of course roughly adjusted by
the rack movement, and finished by the screw, or by such
other means as are chosen for the purpose. One very ingenious contrivance, but applied to the stage, instead of the
body of the microscope, invented by Mr. Powell, will be found
described in the 50th volume of the Transactions of the Society of Arts.

The greater part of the directions for viewing and illuminating objects given in reference to the simple microscope are
applicable to the compound. An argand lamp placed in the
focus of a large detached lens so as to throw parallel rays upon
the mirror, is the best artificial light; and for opaque objects
the light so thrown up may be reflécted by metallic specula
(called, from their inventor, Lieberkhuns) attached to the object-glasses.

It has been recently proposed by Sir David Brewster and by
M. Dujardin to render the Wollaston condenser achromatic,
and they have accordingly been made with three pairs of
achromatic lenses instead of the single lens before described,
with very excellent effect. The last-mentioned gentleman has
also projected an ingenious apparatus, called the Hyptioscope,
attached to the eye-piece for the purpose of erecting the magnified picture.

The erector commonly applied to the compound microscope
consists of a pair of lenses acting like the erecting eye-piece of
the telescope. But this, though it is convenient for the purpose of dissection, very much impairs the op-
tical performance of the instrument.

For drawing the images presented by the
microscope the best apparatus consists of a
mirror M (Fig. 23), composed of a thin piece
of rather dark-colored glass cemented on to a
piece of plate-glass inclined at an angle of 45°
in front of the eye-glass E. The light es-
caping from the eye-glass is assisted in its reflection upwards to the eye by the dark glass,
which effects the further useful purpose of

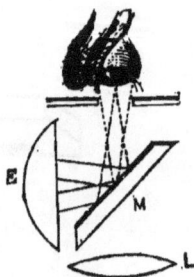

Fig. 23.

rendering the paper less brilliant, and thus enabling the eye better to see the reflected image. The lens L below the reflector

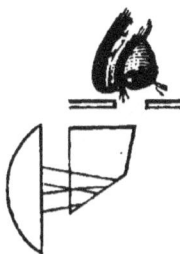

is to cause the light from the paper and pencil to diverge from the same distance as that received from the eye-glass; in other words, to cause it to reach the eye in parallel lines.

Dr. Wollaston's camera lucida, as shown in Fig. 24, is sometimes attached to the eye-piece of the microscope for the same purpose. In this instrument the rays suffer two internal reflections within the glass prism, as will be seen explained in the article "Camera Lucida." In this minute figure we have omitted to trace the reflected rays, merely to avoid confusion.

Fig. 24.

Annexed are four engravings of microscopic objects, the true character of which it is, however, impossible to give in wood, and is difficult indeed to accomplish by any description of engraving.

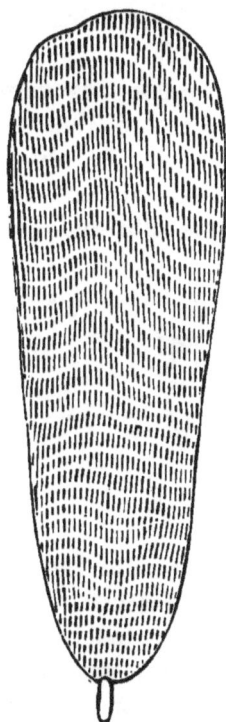

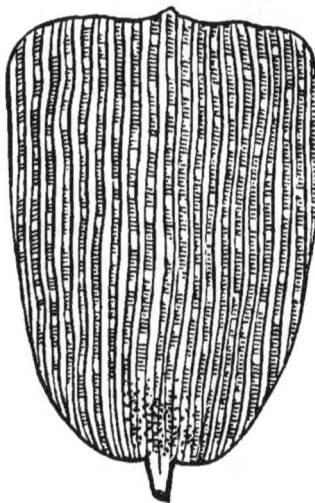

Fig. 25.   Fig. 26.   Fig. 27.

Fig. 25 shows a scale of the small insect called Podura Plumbea, the common Skiptail, magnified about five hundred times. To define the markings on this scale clearly is the highest test of a deep achromatic object-glass; and this drawing is given rather to explain what the observer should look for, than as a very correct representation. Fig. 26 is a scale or feather of the Menelaus Butterfly; Fig. 27 is the hair of a singular insect, the Dermestes; and Fig. 28 is a longitudinal cutting of fir, showing the circular glands on the vessels which distinguish coniferous woods. These latter objects may be seen by half-inch or quarter-inch achromatic glasses. Opaque objects are generally better exhibited by inch and two-inch glasses, when a general view of them is required, and by higher powers when we wish to examine their minute structure. In the latter case the light must be obtained by condensing lenses instead of the metallic specula.

Fig. 28.

Although the reflecting microscope is now very little used, it may be expected that we should mention it. In this instrument, at Fig. 29, the object O is reflected by the inclined face of the mirror M, and the rays are again reflected and converged by the ellipsoidal reflector R R, which effects the same purpose as the object-glass of the compound microscope. It forms an image which is not susceptible of the over-correction as to color before described, and which therefore becomes colored in passing through the eye-piece. This fact, and the loss of light by reflection, will probably always render the reflecting microscope inferior to the achromatic refracting.

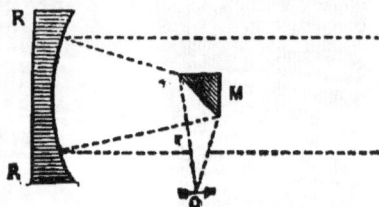

Fig. 29.

The solar microscope has been so nearly superseded by the oxy-hydrogen, that a brief description of the latter must suffice, particularly as their optical principles are similar.

The primary object in both is to throw an intense light upon the object, which is sometimes done by mirrors, and sometimes by lenses. In Fig. 30, L represents the cylinder of burning lime, R R the reflector, which concentrates the light upon the

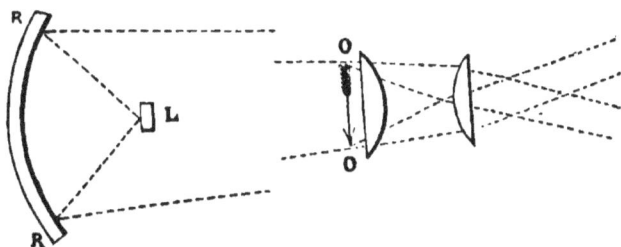

Fig. 30.

object O O; the rays from which, passing through the two plano-convex lenses, are brought to foci upon a screen placed at a great distance, and upon which is formed the magnified image.

Fig. 31 shows a combination of lenses to condense the light upon the object. In either case the optical arrangements by which the image is formed admit of the same perfection as

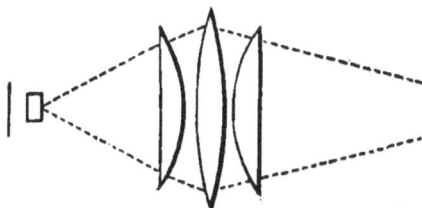

Fig. 31.

those which have been described for the compound micro-scopes. A few achromatic glasses for oxy-hydrogen micro-scopes have been made, and they will ultimately become valu-able instruments for illustrating lectures on natural history and physiology. One made by Mr. Ross was exhibited a few months since at the Society of Arts to illustrate a lecture on the physiology of woods. It should be observed, however, that the oxy-hydrogen or solar microscope requires either a spheri-cal screen, or that the objects should be mounted between spherical glasses, in order to bring the whole into focus at one time. This latter plan was adopted on the occasion just men-tioned with perfect success.